The Woman Celebrant

The Woman Celebrant

Jeremy B. Sims

Published by Jeremy Sims, 2024.

THE WOMAN CELEBRANT

First edition. April 13, 2024.

Copyright © 2024 Jeremy B. Sims.

ISBN: 979-8224631162

Written by Jeremy B. Sims.

Also by Jeremy B. Sims

Book 2
From Country Roads to City Penthouses Part 2

Standalone
Awakened Wellness: Are YOU Spiritually Motivated Yet?
Stop Blaming the Adversary: It's You!
From Milk to Meat: The Journey of Spiritual Maturity
Reviving Your Sacred Space: The Call for True Worshippers
Cultivating Self-Discipline: A Path to Personal and Universal Success
The Morning After
In the Shadow of Holiness
Un Rêve à Washington D.C.
Keep That Temple Clean: A Biblical Perspective
Harboring Harmony: Navigating Through Love, Freedom, and Moral
Rediscovery
and This is just the Beginning
50 Shades of Salvation
Embracing Authenticity
What are YOU fighting for?
Got Humility?
Check your Spam Folder!
Harmony"s Horizon

The Woman Celebrant

Watch for more at https://www.jbsims.com/.

Dedicated to all women, whose daily lives embody resilience, grace, and strength. May this book serve as a celebration of your inherent worth and a reminder that every day, in every way, you are deserving of recognition, respect, and admiration

Introduction

In a world where the achievements and struggles of women are often overshadowed by the prevailing currents of historical narratives and societal norms, "The Celebrant: Women" emerges as a beacon of recognition and appreciation. This book is born out of a desire to illuminate the myriad ways in which women, across the globe and through the ages, have shaped our societies, fought against invisible and visible barriers, and continue to forge paths toward equity and excellence. It is an ode to the resilience, strength, and beauty of womanhood, crafted from a perspective that is seldom foregrounded in discussions about gender equality—the male perspective.

The purpose of this book is multifold. Primarily, it seeks to acknowledge and celebrate the achievements of women—those that have been etched into the annals of history and those that unfold in the quiet corners of everyday life. It is a tribute to the mothers, daughters, leaders, innovators, and dreamers who, in the face of adversity, have persevered to claim their rightful place in the world. This book also aims to shine a light on the struggles that women endure, often silently, offering a platform for these stories to be heard and understood. By weaving together narratives of triumph and trial, "The Celebrant: Women" endeavors to present a holistic view of womanhood, one that is rich with complexity and contradiction.

Understanding and appreciating the essence of womanhood from a male perspective offers a unique lens through which we can explore gender dynamics and the societal constructs that govern them. This viewpoint is not meant to overshadow or co-opt the experiences of women but rather to offer solidarity, empathy, and support. It is an invitation to men to engage deeply with the realities of women's lives, to challenge the patriarchal structures that limit human potential, and to actively participate in the ongoing journey toward gender equality.

"The Celebrant: Women" is more than just a book; it is a call to action. It urges readers, regardless of gender, to recognize the indomitable spirit of women, to celebrate their achievements, and to stand with them in their struggles. Through a blend of historical analysis, personal narratives, and global perspectives, this book seeks to foster a deeper understanding of what it means to be a woman in today's world and the ways in which we can all contribute to a more inclusive, equitable, and celebratory future.

As we embark on this journey together, it is my hope that "The Celebrant: Women" will inspire reflection, conversation, and change. May we all learn to appreciate the beauty of womanhood in all its forms and may we do so with open hearts and minds, ready to be allies in the truest sense of the word.

Chapter 1: The Historical Struggle

The journey of women through the corridors of history is a tapestry woven with the threads of resilience, resistance, and relentless pursuit of equality. From the shadows of obscurity to the limelight of recognition, women have played pivotal roles in shaping societies, cultures, and civilizations. Yet, their contributions have often been marginalized, their struggles invisible. Let's understand the historical roles of women across diverse cultures, examines the constraints that have molded the modern woman, and celebrates the indomitable spirits who defied the norms to etch their names in the annals of history.

Women's Roles Through History Across Different Cultures

The role of women in history varies widely across cultures and epochs, yet common threads of marginalization and resilience run through the tapestry of their experiences. In ancient civilizations, women were often confined to domestic spheres, yet some cultures revered them as goddesses and priestesses, acknowledging their divine connection and wisdom.

ANCIENT CIVILIZATIONS

In ancient civilizations, the societal role of women was predominantly centered around the home and family. In ancient Egypt, women could own property, manage businesses, and were sometimes revered as powerful figures in religion and mythology, like the goddess Isis, embodying the ideals of motherhood and magic. Contrastingly, in ancient Greece, women's roles were significantly more constrained, largely limited to domestic duties, with Spartan women being notable exceptions for their greater freedom and role in managing estates while men were away in battle.

Reverence as Goddesses and Priestesses

Despite these constraints, some cultures elevated women to positions of profound spiritual significance. In various indigenous cultures around the world, women were often seen as the carriers of life and wisdom, deeply connected to the cycles of nature. Celtic societies revered priestesses and female deities, entrusting women with the roles of spiritual guides and guardians of their traditions. Similarly, in Hinduism, goddesses like Saraswati, Lakshmi, and Parvati play crucial roles, representing knowledge, wealth, and power, respectively, reflecting the society's recognition of the divine feminine's power.

Domestic Spheres and Beyond

While the domestic sphere was the primary domain for many women, this did not preclude them from contributing to their societies in significant ways. In many agricultural societies, women played a crucial role in managing food production and household economies. In the Middle Ages, women in Europe ran households, managed estates during their husbands' absences, and participated in crafts and trades. Some, like the famed Hildegard of Bingen, even exceeded these boundaries, making notable contributions to music, medicine, and theology.

Cultures of Equality and Matriarchy

It's also important to highlight cultures that contrasted sharply with patriarchal norms. The Iroquois Confederacy in North America, for instance, was matrilineal, with women holding significant power in decision-making and the selection of leaders. Similarly, in some African societies, women held roles as warriors and leaders, exemplifying strength and governance. These examples showcase the variability in women's roles and the potential for gender equality embedded in different cultural frameworks.

This exploration into the women across various cultures and epochs reveals a complex picture of simultaneous constraint and reverence.

While many societies confined women to domestic roles, others recognized their inherent power, wisdom, and leadership capabilities, either within the spiritual realm as goddesses and priestesses or in the societal sphere as leaders and warriors. The common threads of resilience and marginalization run deep, highlighting the universal struggle of women to claim their space and voice across history. These narratives not only underscore the diversity of women's experiences but also their unwavering strength and the ongoing journey towards achieving gender equity.

Expanding on the historical constraints that have shaped the identity and status of the modern woman requires a closer look at the legal, social, and economic barriers women have faced and continue to navigate. This expanded examination will consider not just the impact on women from a male perspective but will also delve into the interplay between women, examining how these constraints have fostered both competition and camaraderie among them.

LEGAL CONSTRAINTS AND Their Legacy

Historically, legal systems across the world have placed severe restrictions on women's rights and freedoms. For centuries, laws prohibited women from owning property, voting, and pursuing higher education. These legal constraints were not merely external barriers; they were internalized, shaping how women saw themselves and their place in society. For example, the inability to own property or manage finances relegated women to a perpetual state of dependency on male relatives, fundamentally impacting their autonomy and self-esteem.

The legacy of these legal constraints persists, manifesting in modern issues such as the gender wage gap, disparities in property ownership, and unequal representation in political and business leadership positions. While significant progress has been made, with many countries abolishing such discriminatory laws, the echoes of these legal limitations continue to influence societal norms and expectations.

SOCIAL EXPECTATIONS: Femininity, Motherhood, and Beauty

Social norms around femininity, motherhood, and beauty have exerted a profound influence on women's lives, dictating acceptable behavior, appearance, and life choices. These expectations have not only been enforced by men but also, at times, perpetuated and policed by women themselves, creating complex dynamics of conformity and resistance within female communities. The ideal of the "perfect woman" – nurturing, compliant, and beautiful – has been a source of internal conflict for many women, leading to a sense of inadequacy and competition.

However, these same expectations have also fostered a sense of solidarity and mutual support among women. Feminist movements have often arisen from a shared recognition of these oppressive standards, uniting women in their quest for autonomy, respect, and equality. The struggle against restrictive social norms has led to a broader questioning of the very notion of femininity, opening new possibilities for defining identity beyond traditional binaries.

ECONOMIC DISPARITIES and the Path Forward

Economic constraints on women, stemming from both legal restrictions and social norms, have resulted in significant disparities in wealth, employment opportunities, and economic power. Women have historically been steered into lower-paying, less secure forms of employment, contributing to the persistent wage gap and limiting their economic independence. These economic challenges are not only a matter of financial inequality but also impede women's ability to participate fully in public and political life.

The response to these economic disparities has been multifaceted. Women have increasingly entered fields traditionally dominated by men, achieving remarkable successes that challenge stereotypes about women's capabilities and interests. Additionally, women have created supportive networks, providing mentorship, funding, and advocacy for one another in the quest for economic justice and opportunity.

The historical constraints on women—legal, social, and economic—have indeed shaped the identity and status of the modern woman, but they have also catalyzed a powerful response. The struggle against these constraints has led to significant advancements in women's rights and opportunities. Moreover, it has fostered a complex dialogue among women, encompassing both competition and solidarity. As we move forward, understanding these dynamics is crucial for both men and women in continuing to advocate for equality and empowerment for all women.

Key Historical Figures Who Defied the Norms

Delving deeper into the lives and legacies of these remarkable women offers a more nuanced understanding of their contributions, not only in challenging the norms of their times but also in inspiring a broad spectrum of individuals, including men, women, and youth, to pursue equity and justice.

Cleopatra: Power and Diplomacy

Cleopatra VII, the last active ruler of the Ptolemaic Kingdom of Egypt, stands as a testament to female political acumen in an era dominated by male rulers. Her intelligence, political savvy, and strategic use of alliances demonstrate the potential of women to hold power and influence international politics. Cleopatra's reign challenges the traditional narratives of female leadership, providing a powerful example of a woman who navigated the complexities of power dynamics in a highly patriarchal society. Her legacy serves as an inspiration not just to women but to all leaders advocating for smart and strategic governance.

Joan of Arc: Courage and Conviction

Joan of Arc, a peasant girl who led French armies to victory during the Hundred Years' War, embodies the idea that courage and conviction can break the shackles of gender norms. Her story is a powerful reminder that leadership and heroism are not confined to one gender. Joan challenged the societal expectations of her time, not only regarding women's roles in warfare but also in how women could influence the course of history. Her unyielding spirit and bravery inspire not only women but also men and young people to fight for their beliefs, even in the face of overwhelming opposition.

Sojourner Truth: Advocacy for Equality

Sojourner Truth, born into slavery, emerged as a formidable advocate for abolition and women's rights. Her famous "Ain't I a Woman?" speech brilliantly highlighted the intersections of racial and gender injustices, challenging both white women to recognize their racial privileges and men of all races to acknowledge their gender privileges. Truth's activism underscores the importance of inclusive equality movements that address the multifaceted nature of discrimination. Her legacy is a beacon for activists today, encouraging a unified approach to fighting for gender and racial equality that includes everyone, regardless of their background.

Marie Curie: Breaking Scientific Barriers

Marie Curie, the first woman to win a Nobel Prize and the only person to win in two different scientific fields, shattered the glass ceiling in the world of science. Her groundbreaking work in physics and chemistry was achieved in an environment that was not only male-dominated but often openly hostile to women. Curie's perseverance and unparalleled contributions to science have opened doors for countless women in STEM fields, inspiring both women and men to pursue their passion for science, irrespective of gender-based obstacles.

Malala Yousafzai: A Voice for Education

Malala Yousafzai, a Pakistani activist for female education, became a global symbol of resistance against the Taliban's efforts to deny girls an education. Surviving an assassination attempt, Malala's advocacy highlights the power of youth voice in championing human rights. Her story is a potent reminder of the ongoing struggles for education rights in many parts of the world and serves as an inspiration not only to young girls but also to boys and adults, emphasizing the role of education in achieving gender equality and empowering future generations.

The stories of these women, spanning different cultures and centuries, illustrate the myriad ways in which women have resisted oppression and championed change. Their legacies are not confined to their achievements in breaking gender barriers but extend to their role in shaping a more inclusive world. These women have paved the way not only for future generations of women but have also set a precedent for men and youth, showing that the fight for equality requires the support and involvement of all. Their lives encourage ongoing dialogue and action across genders and generations, fostering a world where everyone has the opportunity to contribute to societal progress.

Chapter 2: Global Perspectives on Womanhood

Exploration of Womanhood in Various Cultures

Diverse Cultural Interpretations Womanhood carries distinct meanings across different cultures, shaped by historical, social, and spiritual narratives. In some societies, womanhood is deeply intertwined with traditional roles of motherhood and caregiving, while in others, it encompasses the pursuit of education, career, and independence alongside domestic responsibilities. For instance, in many Indigenous cultures, womanhood is revered, with women often seen as the heart of communities, embodying strength, wisdom, and a deep connection to nature and spirituality. In contrast, Western societies might emphasize individual achievement and self-expression as key components of womanhood.

Food for thought

To elevate society beyond restrictive perspectives on womanhood and enhance women's ability to choose and define their paths according to their specific needs, a multifaceted approach is necessary. This approach should involve cultural, educational, policy-driven, and community-based strategies to foster an environment where every woman has the autonomy to shape her own identity and future. Here's how this can be achieved:

Cultural Shifts

Promoting Diverse Narratives: Society must embrace and promote a diversity of narratives around womanhood through media, literature, and the arts. By showcasing a wide range of women's stories and experiences, we can challenge and expand the prevailing notions of what it means to be a woman, making it clear that there is no single correct way to embody womanhood.

Encouraging Male Allies: It's crucial to engage men and boys as allies in redefining gender norms. Educational programs and campaigns that encourage empathy, respect, and equality can help dismantle harmful stereotypes, fostering a culture where everyone supports each other's choices and freedoms.

Educational Empowerment

Comprehensive Gender Education: Implementing comprehensive gender education in schools can challenge traditional gender roles and stereotypes from an early age. Education should emphasize the value of diversity, the importance of equality, and the rights of individuals to pursue their paths without being confined by societal expectations.

Career and Leadership Opportunities: Providing girls and women with more opportunities to explore a variety of careers, including those in STEM fields, leadership, and areas traditionally dominated by men, empowers them to make informed choices about their futures. Mentorship programs, scholarships, and workshops can play a significant role in opening these doors.

Policy and Legal Frameworks

Legislative Support for Gender Equality: Enacting and enforcing laws that guarantee equal rights and opportunities for women in all aspects of life—education, employment, healthcare, and political representation—is fundamental. Policies should also address the wage gap, maternity and paternity leave, and protection against all forms of violence.

Support Systems for Women: Developing robust support systems for women, including access to healthcare, childcare, and social services, can significantly impact their ability to make autonomous choices. These systems should be designed to meet diverse needs and be accessible to all women, regardless of their socio-economic status.

Community Engagement

Building Supportive Communities: Communities play a key role in supporting women's choices. Initiatives that foster networking, mentorship, and solidarity among women can empower them to pursue their goals. Community centers, online platforms, and local organizations can offer resources and support for women navigating their personal and professional lives.

Celebrating Women's Achievements: Regularly celebrating the achievements of women in various fields can inspire others and reinforce the message that women can excel in any area they choose. Awards, public recognition, and media spotlights on successful women can help shift societal perspectives on womanhood.

Elevating society's view of womanhood to enhance women's ability to choose their paths requires a collective effort across cultural, educational, policy, and community spheres. By actively working to shift narratives, empower through education, support with policy, and engage communities, we can create a more inclusive and equitable society where every woman has the autonomy to determine her life's direction, free from the constraints of traditional gender roles. This approach not only benefits women but enriches society by embracing and nurturing a diversity of talents, perspectives, and potentials.

Honoring Diverse Expressions To honor these diverse expressions of womanhood, it is crucial to support cultural initiatives that celebrate these various identities, including festivals, art exhibits, and educational programs that highlight the richness of women's roles in different societies.

Challenges Faced by Women Globally

Economic Disparities Globally, women face wage gaps, underrepresentation in leadership roles, and higher rates of economic insecurity. In developing countries, women often have limited access to financial services and property rights, which hampers their ability to build wealth and invest in their futures.

Social and Political Barriers Social norms and political structures frequently limit women's voices and choices. From gender stereotypes that pigeonhole women into certain roles to laws that restrict their freedoms and participation in public life, these barriers are pervasive. Additionally, violence against women remains a critical issue worldwide, significantly impacting their ability to live freely and safely.

To assist in supporting women's economic empowerment and promoting gender equality effectively, it's beneficial to break down the approach into actionable, digestible chunks. Each element plays a critical role in addressing the multifaceted challenges women face globally. Here's how these efforts can be organized and implemented:

Economic Empowerment

1. Access to Education:

- **Initiatives:** Scholarships, online courses, and women-centric educational programs designed to increase access to primary, secondary, and higher education for women and girls.
- **Impact:** Education is foundational in empowering women to pursue higher-paying jobs and leadership roles, breaking cycles of poverty and dependency.

2. Vocational Training:

- **Initiatives:** Skills development programs in trades, technology, and entrepreneurship tailored specifically for women, including workshops and apprenticeships.
- **Impact:** Vocational training equips women with marketable

skills, enhancing their employability and ability to start their own businesses.

3. Microfinance and Economic Support:

- **Initiatives:** Microloans, savings clubs, and financial literacy programs targeted at women, especially in underserved communities.
- **Impact:** Access to credit and financial services enables women to invest in their entrepreneurial ventures, supporting economic independence and community development.

Advocacy for Policy Change

1. Promoting Gender Equality Policies:

- **Action:** Lobbying for the creation and enforcement of laws that ensure equal pay, protect against discrimination in the workplace, and support women's health and rights.
- **Impact:** Strong legal frameworks protect women's rights and ensure that they have equal opportunities in all aspects of society.

2. Protection Against Violence:

- **Action:** Advocating for comprehensive laws and services to prevent gender-based violence, support survivors, and penalize perpetrators.
- **Impact:** A safer environment for women enhances their freedom to participate in economic, social, and political life.

Political and Social Participation

1. Supporting Female Candidates and Leaders:

- **Action:** Campaigns and funding support for women running for office, leadership training programs, and initiatives to increase the visibility of female leaders.
- **Impact:** Increased representation of women in politics and leadership positions can lead to more inclusive and equitable decision-making processes.

2. Promoting Gender-Sensitive Policies:

- **Action:** Encouraging the integration of gender perspectives in policymaking, ensuring that policies address the specific needs and challenges of women and men equally.
- **Impact:** Gender-sensitive policies contribute to a more equitable society by recognizing and addressing the different impacts of decisions on women and men.

Implementation Strategies

- **Community Engagement:** Grassroots movements and community groups are crucial for raising awareness and implementing local initiatives that support women's empowerment.
- **Partnerships:** Collaborating with NGOs, governments, and the private sector can amplify the impact of initiatives aimed at supporting women.

- **Awareness Campaigns:** Using media and social platforms to educate the public on the importance of gender equality and the specific needs of women in their communities.

By breaking down the approach into these focused areas, individuals and organizations can more effectively contribute to the overarching goal of empowering women economically, protecting their rights, and ensuring their participation in all spheres of life. This structured approach allows for targeted actions, making it easier to track progress and measure the impact of these initiatives.

Success Stories of Women Overcoming Adversity

Global Inspirations Women like Ellen Johnson Sirleaf, the first elected female head of state in Africa, who led Liberia through reconciliation and recovery; Malala Yousafzai, who became a global advocate for girls' education after surviving an assassination attempt; and Greta Thunberg, who has inspired an international movement to combat climate change, are testament to the incredible resilience and leadership of women facing and overcoming adversity.

Celebrating and Supporting Success Celebrating these successes involves not only acknowledging these women's achievements through media and educational curricula but also supporting the causes they champion. Engaging with and contributing to organizations and movements led by women, providing platforms for their stories to be heard, and mentoring young women to follow in their footsteps are all ways to honor and extend their legacies.

Global perspectives on womanhood reveal a complex interplay between cultural heritage, societal challenges, and individual triumphs. By understanding and honoring the diverse expressions of womanhood, actively working to dismantle the barriers women face, and celebrating the achievements of those who overcome adversity, we can all contribute to a more equitable and just world. This insert not only highlights the struggles and successes of women globally but also serves as a call to action for everyone to support the journey towards full gender equality.

Chapter 3: The Psychological Journey

The psychological journey of women is marked by a unique set of challenges and experiences, deeply influenced by societal norms, cultural expectations, and personal struggles. This chapter delves into the psychological hurdles specifically encountered by women, such as body image issues and mental health concerns, examines how societal pressures shape women's self-perception and mental well-being, and explores strategies for fostering resilience and self-acceptance.

Psychological Challenges Unique to Women

Body Image Issues:

Women are often subjected to unrealistic beauty standards that are pervasive in media and cultural narratives. This constant exposure can lead to a range of body image issues, from negative self-perception to severe disorders like anorexia and bulimia.

Impact: These issues can severely affect women's mental health, leading to depression, anxiety, and low self-esteem.

Mental Health:

Women are at a higher risk for certain mental health conditions, such as depression and anxiety, partly due to hormonal factors, but also because of the cumulative stress of juggling multiple roles in society.

Compounding Factors: Discrimination, violence, and societal pressures can exacerbate mental health struggles, making it harder for women to seek and receive help.

The Impact of Societal Expectations

On Self-Perception:

Societal expectations often dictate how women should look, behave, and even think. This can create a disconnect between one's true self and the self that society demands, leading to internal conflict and distress.

Consequences: The pressure to conform to these expectations can erode women's self-confidence and authenticity, making it difficult for them to value their own strengths and qualities.

ON MENTAL HEALTH:

The relentless pursuit to meet societal expectations can be mentally exhausting and isolating, contributing to stress, anxiety, and depression.

Intersectionality: For women of color, LGBTQ+ women, and women with disabilities, these expectations are further complicated by additional layers of stereotypes and biases, intensifying the psychological toll.

Strategies for Resilience and Self-Acceptance

Building Resilience: Resilience can be fostered through supportive networks, whether they are found in friendships, family, or communities that affirm women's worth beyond societal standards.

Practices: Engaging in mindfulness, therapy, and self-care routines can equip women with the tools to manage stress and navigate challenges

Promoting Self-Acceptance:

Encouraging women to embrace their individuality and reject unrealistic standards is crucial. This can involve media literacy programs to critically assess the images and messages that shape beauty standards.

Empowerment: Empowering women through education, leadership opportunities, and platforms to share their stories can reinforce the value of diverse experiences and bodies.

The psychological journey for women is fraught with challenges uniquely shaped by the intersection of gender and societal expectations. Addressing these challenges requires a concerted effort to change societal norms, provide support systems, and empower women to embrace their true selves. By fostering an environment that celebrates diversity, resilience, and self-acceptance, we can contribute to the psychological well-being of women and support their journey towards a more fulfilling life.

Chapter 4: Celebrating Womanhood in Daily Life

Acknowledging and celebrating the achievements of women in daily life is essential for shifting the dynamics of a patriarchal society towards one of equality and mutual respect. Let's highlight the often-overlooked accomplishments of women, stress the importance of recognizing their daily victories, and provide insights into how men can better appreciate and support the women in their lives. The goal is to speak directly to men in a clear, relatable manner, encouraging a deeper understanding and active participation in celebrating womanhood.

(Now pass this section to the man or men in your life)

———

R*ecognizing Everyday Achievements*

The Unseen Work:

Think about the women around you—their day-to-day involves a lot more than meets the eye. From managing households to juggling careers and family responsibilities, women often perform a balancing act that goes largely unnoticed.

Action Point for Men: Start noticing the small stuff. Did a female colleague lead a project to success? Did your partner manage to calm a tantrum and finish a work call simultaneously? Acknowledge it. A simple "I see what you did there, and it's impressive" can go a long way.

Celebrating Small Victories

Why It Matters:

In the grand scheme of things, it's the small victories that add up to significant achievements. Celebrating these moments can boost confidence and morale, and it's crucial for women to feel seen and appreciated in their daily efforts.

Action Point for Men: Did she finish a presentation that she's been stressing over? Maybe she fixed something around the house or helped your kid with a tough homework assignment. Say it out loud: "That's really awesome; I'm proud of you." It's about recognizing effort and the achievement.

Male Perspectives on Appreciation

Understanding and Empathy:

To truly appreciate the women in our lives, we, as men, need to step into their shoes and understand the unique challenges they face. It's about empathy—recognizing that even if we don't experience the same pressures, they are very real for the women around us.

Action Point for Men: Listen more. Before offering solutions, just listen to her day, her struggles, and her victories. Sometimes, being heard is all one needs to feel appreciated.

Active Support:

Appreciation goes beyond words. It's shown through actions. Supporting the women in our lives means sharing responsibilities, standing up for them, and actively contributing to their success and well-being.

Action Point for Men: Get involved. Help without being asked, whether it's with chores, childcare, or just being there emotionally. And when she achieves something, be her biggest cheerleader.

For men, celebrating womanhood in daily life is about recognizing the unseen, acknowledging the effort, and showing appreciation and support. It's a straightforward but powerful way to contribute to healing and changing the dynamics of a patriarchal society. By valuing the everyday achievements of women and actively participating in their recognition, we pave the way for a more inclusive, respectful, and equitable world. Remember, it's the little things that count, and as men, our role in this journey is crucial.

Chapter 5: Women Against Women: Transforming Rivalry into Solidarity (Difficult conversations for women)

In the tapestry of gender equality and women's empowerment, a less discussed but significant issue is the phenomenon of women against women. Look into the complexities of female rivalry, its roots in societal structures, and the explicit need for a shift towards solidarity and mutual support among women. Understanding and addressing this dynamic is crucial for advancing women's collective interests in both personal and professional spheres.

Understanding the Roots of Female Rivalry

Female rivalry, often characterized by competition, jealousy, and conflict among women, is not an inherent trait but a manifestation of patriarchal societal norms that pit women against each other. From a young age, girls are subjected to comparisons based on appearance, behavior, and later, professional achievements, fostering a sense of competition rather than camaraderie. The scarcity of leadership roles for women and societal expectations about femininity further exacerbate this issue, creating environments where women may feel they must compete for limited resources and recognition.

THE IMPACT OF RIVALRY on Women's Progress

The consequences of women against women extend beyond individual relationships, hindering collective progress towards gender equality. When women view each other as competitors rather than allies, it diminishes the potential for collective action and support networks that are vital for overcoming systemic barriers. This rivalry not only affects mental and emotional well-being but also obstructs opportunities for mentorship, collaboration, and shared success.

Shifting the Paradigm: From Rivalry to Solidarity

The transformation from rivalry to solidarity among women requires both individual and collective efforts. Recognizing the external pressures that fuel competition and consciously choosing collaboration over competition are initial steps toward change. Encouraging environments that celebrate diverse achievements and foster open communication can help dismantle the notion that one woman's success is another's failure.

FOSTERING FEMALE MENTORSHIP and Support Networks

Creating and nurturing female mentorship programs and support networks is pivotal in changing the narrative. By sharing knowledge, resources, and opportunities, women can uplift each other, creating a multiplier effect that benefits all involved. These networks not only provide professional and personal support but also serve as powerful platforms for advocating gender equality and challenging societal norms.

Celebrating Collective Success

Acknowledging and celebrating the achievements of women as collective victories can significantly impact the perception of female relationships. Highlighting stories of collaboration, mutual support, and shared triumphs in various fields reinforces the idea that solidarity among women is not only possible but also immensely rewarding.

The journey from women against women to a culture of solidarity and mutual support is essential for the advancement of gender equality. By examining the roots of female rivalry, recognizing its detrimental effects, and actively working towards fostering an environment of collaboration and support, women can break the cycle of competition. This underscores the importance of solidarity in achieving collective goals and transforming societal norms, highlighting that when women come together, their potential is boundless.

Chapter 6: Shifting the Paradigm: From Rivalry to Solidarity

Understanding Historical Backdrop

The competition among women, often seen as a rivalry, has deep historical roots intertwined with societal norms and gender expectations. Historically, women have been relegated to roles that emphasize their relationships to men—daughters, wives, mothers—rather than individuals with their own ambitions and talents. Opportunities for women to engage in public, economic, and intellectual life were scarce, creating a zero-sum game where the success of one woman could seemingly come at the expense of another. This backdrop has perpetuated a culture where women are often pitted against each other, competing for limited spaces in professional settings, social recognition, and even in personal relationships.

In many cultures, women's worth was often measured in comparison to other women, whether through beauty, domestic ability, or moral virtue, further fueling competition rather than camaraderie. This historical context has laid the foundation for modern challenges, where despite significant progress in gender equality, remnants of these competitive dynamics persist.

LOGIC FOR ENHANCING Camaraderie

1. **Breaking the Cycle of Scarcity:** The first step in enhancing camaraderie among women involves dismantling the myth of scarcity—that there are limited spots for women at the table. This myth feeds into the competitive mindset. Recognizing that one woman's success does not detract from another's but rather contributes to the overall advancement of women is crucial. By fostering an abundance mindset, women can view each other as allies in a collective struggle against systemic

barriers.

2. **Celebrating Diversity and Intersectionality:** Embracing and celebrating the diversity among women—acknowledging different backgrounds, experiences, and challenges—can enhance camaraderie. This involves recognizing the unique obstacles faced by women of color, LGBTQ+ women, women with disabilities, and others. Intersectionality provides a framework for understanding how various forms of inequality and discrimination intersect and emphasizes the importance of supporting all women, not just those who share our own experiences or backgrounds.

3. **Creating Spaces for Open Communication:** Encouraging environments that prioritize open, honest communication allows women to share their experiences, challenges, and successes. These spaces can be physical or virtual, formal or informal, but the key is that they provide a platform for dialogue and understanding. Through these conversations, women can dismantle stereotypes, dispel myths of rivalry, and build empathy and mutual respect.

4. **Mentorship and Support Networks:** Establishing mentorship programs and support networks among women can play a pivotal role in shifting from rivalry to solidarity. These initiatives provide opportunities for women to learn from each other, offer guidance, and extend support. Mentorship and networks not only help individual women advance their careers but also foster a culture of giving back and lifting others as they climb.

5. **Collective Action and Advocacy:** Engaging in collective action and advocacy for gender equality is a powerful way to enhance camaraderie among women. When women come together to fight for common causes—such as equal pay, reproductive rights, and against gender-based violence—they

build a sense of solidarity and shared purpose. These movements underscore the idea that together, women can effect change on a larger scale than any individual could alone.

The shift from rivalry to solidarity among women is not merely beneficial but essential for the continued progress towards gender equality. By understanding the historical backdrop of competition and embracing the logic for enhancing camaraderie, women can collectively dismantle systemic barriers and create a more inclusive, equitable future. This transformation requires conscious effort, collective action, and a commitment to supporting one another in all aspects of life.

Chapter 7: A Call to Action: Women Uniting for Change

To every woman, across all cultures, of every identity and pronoun,

It's time to stand at a pivotal moment in history, a juncture where collective action can usher in a transformative era of equality, inclusivity, and empowerment. The journey ahead, laden with the potential to dismantle systemic barriers that have long hindered progress, requires more than individual effort—it demands unity, solidarity, and a shared vision for a better future.

Embrace the Power of Collaboration

The beauty and strength of women diversity are unmatched. When you come together, recognizing, and valuing the unique experiences and perspectives each of you bring, you create an unstoppable force for change. It's time to shed the outdated notions of rivalry and competition that have been imposed on women. Instead, embrace the power of collaboration. Together, you can achieve more than you ever could alone.

Acknowledge and Celebrate the Differences

Women differences are not just to be acknowledged—they are to be celebrated. Each woman's story adds a thread to the rich tapestry of the collective experience. By embracing intersectionality, you ensure that no one is left behind, and every struggle and success is recognized. It's through this acknowledgment that you can address the unique challenges faced by women of color, LGBTQ+ women, women with disabilities, and others who have been marginalized within the communities.

Foster Environments of Open Communication and Support

Commit to creating spaces where every woman can speak her truth, share her journey, and find support. These environments—whether they're in your homes, workplaces, or online communities—are the foundations upon which you build understanding, empathy, and solidarity. Open, honest communication is the tool for breaking down barriers and building bridges.

Engage in Collective Action

The challenges you face are not insurmountable. Together, through collective action and advocacy, you can tackle the issues that affect all women —from fighting for equal pay and reproductive rights to combating gender-based violence. Women unity has the power to enact real, lasting change, not just for yourselves but for future generations.

COMMIT TO LIFTING EACH Other Up

As you move forward, start making a conscious commitment to support one another in all aspects of life. Start to be a mentor, guide, and uplift each other. Celebrate every victory, no matter how small, and stand together in every struggle. It's through this unwavering support and solidarity that you can overcome the obstacles that lie in your path.

A Harmonious Global Impact

Imagine a world where women, in all their diversity, stand together in solidarity. A world where collaborative efforts bring about a harmonious global impact, breaking down the walls of inequality and creating a future where everyone, regardless of gender or pronoun, could thrive. This is the world you can build—a world of equality, justice, and endless possibility.

Let this be the call to action if you have not started a call on your own. Together, create a more inclusive, equitable future for all women. The time is now, and every one of you has a role to play. Start embracing the journey ahead with open hearts and united spirits, for in solidarity, there is strength, and in unity, there is power.

To all women, everywhere: the future is yours to shape. Do it together and become the representation of women in leadership.

Chapter 8: Women and Leadership

In today's modern society, the landscape of leadership across various sectors—be it in corporate, political, or community settings—reveals a stark underrepresentation of women. This discrepancy not only highlights a significant loss in diversity and perspective but also underscores a broader issue of systemic barriers that women face, which are further compounded by differences in religion, culture, and socioeconomic status. Despite these obstacles, many women have not only navigated these challenges with resilience and determination but have also significantly contributed to reshaping the leadership paradigm for future generations.

Women in leadership roles often encounter a multifaceted set of challenges, magnified by societal expectations and institutional biases. These obstacles range from gender stereotypes that question their authority and competence, to a lack of access to crucial networks and mentorship opportunities that are readily available to their male counterparts. For women of diverse religious backgrounds, these challenges can be further intensified by cultural norms and expectations specific to their faith communities, which may restrict their participation in public life or impose additional roles and responsibilities that hinder their professional advancement.

Moreover, the balancing act between professional aspirations and personal commitments poses a significant hurdle for many women aspiring to leadership positions. This challenge is universal but can vary in intensity across different cultures and religious contexts. For example, in some communities, women are expected to prioritize family and domestic responsibilities, making it difficult for them to pursue career advancement and leadership opportunities. The societal expectation for women to seamlessly manage both spheres often leads to a phenomenon known as the "double bind," where women are penalized for not meeting high standards in both their professional and personal lives.

Despite these formidable challenges, countless women across the globe have broken through barriers to assume leadership roles, demonstrating that it is possible to redefine what it means to be a leader in today's society. These pioneering women have employed a range of strategies to overcome obstacles, from leveraging supportive networks and finding mentors who champion their advancement, to advocating for policy changes within their organizations and industries that foster gender diversity and inclusion. Their resilience and success in the face of adversity serve not only as a testament to their individual capabilities but also as a source of inspiration for other women navigating similar paths.

Celebrating female leaders who have paved the way is crucial in highlighting the diverse capabilities and perspectives women bring to leadership roles. These trailblazers come from all walks of life, encompassing various religious backgrounds, ethnicities, and classes, illustrating that leadership knows no bounds. By honoring their achievements and the challenges they've overcome, we not only acknowledge their contributions but also underscore the importance of creating more inclusive environments that enable women from all backgrounds to rise to leadership positions.

The stories of these women resonate deeply with readers, offering both a mirror to reflect on the societal structures that continue to hinder women's progress and a window into the possibilities that lie ahead. As we move forward, it is essential to continue challenging the status quo, supporting policy reforms, and cultivating environments that recognize and nurture the leadership potential of all women. By doing so, we not only advance gender equality but also enrich our societies with diverse leadership that can navigate the complexities of the modern world with empathy, innovation, and inclusivity.

Chapter 9: The Power of Female Networks

In the quest for gender equality and the empowerment of women, the role of support networks and communities cannot be overstated. These networks serve as the backbone for encouragement, guidance, and the sharing of invaluable resources and experiences. From local community groups to global online forums, these spaces offer women and girls a platform to connect, learn, and grow. They reinforce the notion that no one is alone in their struggles or ambitions, providing a collective strength that can propel individuals to achieve remarkable feats.

Success stories of women uplifting women abound, showcasing the incredible impact of mutual support. Consider the story of a young girl who, inspired by a women's coding club in her community, pursued her passion for technology despite facing discouragement due to her gender. With mentorship and support from women within the network, she not only excelled in her studies but also went on to inspire other girls to follow in her footsteps, creating a ripple effect of empowerment. Such stories highlight the transformative power of women supporting each other, breaking barriers, and paving the way for future generations.

Men play a crucial role in these support networks as well. Their involvement can range from advocating for gender equality within their own circles to actively participating in initiatives that empower women. By mentoring, sharing responsibilities equally at home and work, and standing against gender biases, men contribute to fostering an environment where women and girls can thrive. Encouraging young boys to engage in conversations about gender equality and respect from an early age is essential. It helps in cultivating a culture of mutual respect and support, ensuring that as they grow, they become allies in the movement towards gender equality.

Creating relatable and inclusive support networks that welcome women and girls from all walks of life is vital. Such environments not only empower individuals but also enrich communities by harnessing the diverse talents and perspectives of women. By nurturing these networks, society can ensure that every girl growing into womanhood knows she has a community behind her, ready to uplift her ambitions and support her journey. It's in these shared spaces of encouragement and collaboration that women, and indeed society, find the strength to challenge the status quo and create a more equitable world for all.

Chapter 10: Holistic Health and Womanhood

The well-being of women encompasses much more than the absence of illness; it is a holistic balance of physical, mental, and spiritual health. This comprehensive approach to health is vital for women of all ages, from young girls blossoming into womanhood to those navigating the complexities of adulthood and beyond. Physical health lays the foundation, enabling women to engage actively in their lives and pursuits. Regular exercise, a nutritious diet, and adequate rest are crucial components, but so is access to healthcare that addresses the unique medical needs of women at different stages of their lives.

Mental health, equally important, involves nurturing emotional well-being and managing stress through practices like mindfulness, therapy, and community support. The societal pressures and gender-specific challenges women face can take a toll on their mental health, making it essential to cultivate resilience and seek support when needed. Spiritual health, though sometimes overlooked, plays a significant role in providing a sense of purpose and connection. Practices such as meditation, spending time in nature, or engaging in activities that align with personal beliefs and values can enhance spiritual well-being, contributing to a more fulfilled and balanced life.

Holistic health practices, such as yoga (as one would like to choose) , meditation, and the use of natural remedies, offer numerous benefits by integrating physical, mental, and spiritual health into a cohesive approach to well-being. These practices encourage mindfulness, reduce stress, and improve physical fitness, addressing health from a multi-dimensional perspective that is especially beneficial for women. They provide tools for women to take charge of their health, empowering them to live harmoniously and resiliently.

The role of male acknowledgment and support in women's health is indispensable. When men understand and empathize with the unique health challenges women face, they can become allies in advocating for women's health rights and supporting their well-being. This support can manifest in many ways, from participating in shared health activities to encouraging and facilitating access to healthcare services. By fostering an environment where women's health needs are openly acknowledged and prioritized, society can ensure that women and girls receive the care and support necessary to thrive at every stage of their lives.

Engaging young girls in conversations about holistic health and well-being from an early age is crucial. It equips them with the knowledge and tools to prioritize their health as they grow, ensuring they enter womanhood empowered to take care of themselves in all aspects. As society embraces a more inclusive approach to health, recognizing the interconnectedness of physical, mental, and spiritual well-being, women across the spectrum can achieve a healthier, more balanced life, supported by the men in their lives and the broader community.

Chapter 11: Embracing Femininity and Strength

The myth that femininity equates to weakness is a deeply ingrained stereotype that has persisted across cultures and generations. However, this notion fails to capture the complexity and strength inherent in femininity. Femininity, in its essence, encompasses a range of qualities and attributes that are not only diverse but also powerful. The resilience, empathy, and nurturing instincts often associated with femininity are, in fact, sources of profound strength. These traits enable women to lead, innovate, and inspire change in unique and impactful ways.

Celebrating the strength in vulnerability and emotional expression is crucial in redefining societal perceptions of femininity. Vulnerability is not a sign of weakness; it is a courageous act of opening oneself to others, fostering genuine connections, and embracing the full spectrum of human emotions. Emotional expression, similarly, allows for the communication of needs, desires, and boundaries, creating deeper understanding and empathy within relationships and communities. These aspects of femininity enrich our social fabric, encouraging a more compassionate and connected world.

Male perspectives play a significant role in dismantling stereotypes about femininity. When men recognize and respect the strength in femininity, they contribute to a cultural shift towards valuing and celebrating these qualities in everyone, regardless of gender. Understanding femininity not as a limitation but as a vital component of human experience allows men to appreciate the diverse expressions of strength around them. This understanding fosters healthier relationships and communities where everyone's contributions are valued and respected.

Men can actively participate in this cultural shift by challenging their own perceptions, engaging in open dialogues about gender stereotypes, and advocating for environments where femininity is embraced and respected. By doing so, men not only become allies in the fight against gender-based stereotypes but also unlock the potential to explore their own relationship with vulnerability and emotional expression, leading to richer, more fulfilling lives.

Debunking the myth that femininity equates to weakness is essential for creating a more equitable society. Celebrating the inherent strength in femininity, recognizing the power of vulnerability and emotional expression, and encouraging men to understand and respect these qualities, we can collectively foster a world that honors and uplifts the full range of human experience, making space for everyone to thrive.

Chapter 12: The Future of Womanhood

Envisioning the future of gender equality involves imagining a world where the roles and contributions of women are recognized and valued equally across all spheres of society. In this future, gender no longer dictates one's path in life, capabilities, or rights. Women and men work side by side, with equal opportunities to succeed and lead. This vision is not just a dream; it's a goal that society is steadily working towards. The realization of this future hinges on dismantling systemic barriers, challenging outdated norms, and fostering an environment where every individual, regardless of gender, can thrive.

The cornerstone of this future is education and opportunity. Access to quality education for girls and young women is crucial, as it lays the foundation for their empowerment and success. Education opens doors to new possibilities, providing the tools needed to challenge inequalities and break cycles of poverty. Beyond primary education, opportunities for higher learning and professional development must be equally accessible to women, allowing them to excel in fields traditionally dominated by men. By investing in the education and empowerment of the next generation of women, society ensures a future workforce, leadership, and innovation that are truly representative of its diversity.

Men play a critical role as allies in the journey towards gender equality. Allyship involves recognizing the privileges afforded by a patriarchal society and actively working to dismantle those systems of inequality. Men can support the women in their lives through listening, learning, and advocating for change. This means challenging sexist remarks and behaviors, supporting policies that promote gender equality, and sharing responsibilities equally in both the workplace and the home. By doing so, men contribute to creating a culture that values and respects women as equals.

Men's involvement in the fight for gender equality is not just about making space for women; it's about reshaping society to benefit everyone. When men stand as allies, they help to break down the harmful stereotypes that limit both women and men. The journey towards gender equality is enriched by the perspectives and efforts of people of all genders, working together to create a more inclusive, equitable world.

As time draws near, the future of gender equality is bright, built on the pillars of education, opportunity, and allyship. By nurturing the potential of the next generation of women, challenging societal norms, and fostering meaningful partnerships across genders, society can move closer to realizing a world where everyone has the freedom to pursue their aspirations without the constraints of gender bias. This future is not only possible; it is within reach, with every act of support and solidarity moving us one step closer.

Conclusion

Summary and Call to Action

Throughout this book, we've journeyed through the diverse experiences of womanhood, from the historical struggles and achievements of women across cultures to the importance of fostering physical, mental, and spiritual well-being. We've examined the systemic barriers that women face and celebrated the resilience and solidarity that mark the path towards gender equality. The key themes—empowerment through education, the dismantling of stereotypes, the celebration of femininity, and the critical role of support networks—underscore the multifaceted nature of womanhood and the collective effort required to uplift and support women in all aspects of life.

Call to Action

This exploration culminates in a call to action for everyone, regardless of gender, to actively participate in the celebration and support of womanhood. For women, this means embracing your journey, recognizing your strength, and supporting your sisters in their endeavors. For men, it involves acknowledging the unique challenges women face, educating yourselves on gender equality, and standing as allies alongside women.

Celebrate Womanhood

Celebrating womanhood involves recognizing and honoring the achievements, both big and small of women in our lives. It's about creating spaces where women feel seen, heard, and valued. This celebration is not just for one day or month: it's a continuous acknowledgment of the contributions and resilience of women.

Support Womanhood

Supporting womanhood means advocating for policies and practices that promote gender equality and protect women's rights. It's about challenging the status quo, questioning gender norms, and dismantling the barriers that hinder women's progress. Support can take many forms, from mentorship and advocacy to sharing household responsibilities equally and respecting women's choices and boundaries.

A Unified Effort

The journey towards gender equality and the empowerment of women is a shared responsibility. It requires the collective action of individuals, communities, and societies at large. By standing together, celebrating the achievements of women, and addressing the challenges they face, we can create a more inclusive, equitable world for future generations.

Embrace the Future

As we look towards the future, let's commit to being agents of change in our communities. Let's educate ourselves and others, advocate for equality, and support one another in our aspirations and struggles. Together, we can break down the barriers that divide us and build a world where everyone, regardless of gender, could thrive.

Reflecting on the journey of understanding and appreciating the essence of being a woman reveals a rich tapestry of experiences, struggles, and triumphs that define womanhood. This journey is not just about recognizing the challenges women face across different cultures and epochs but also about celebrating the resilience, strength, and diversity that women embody. Through this exploration, we've delved into the historical context of women's roles, the barriers that have shaped their experiences, and the victories, both large and small, that mark their path towards empowerment.

APPRECIATING THE ESSENCE of being a woman involves acknowledging the complexities and nuances of their experiences. It's about seeing beyond the stereotypes and recognizing the individual and collective strengths that women bring to the table. From the courage of women who have broken barriers in leadership and innovation to the everyday acts of resilience in the face of adversity, each story adds depth to our understanding of what it means to be a woman.

This journey also highlights the importance of solidarity among women and the pivotal role of supportive networks and communities in empowering women to realize their full potential. It underscores the need for an inclusive approach to feminism, one that embraces the diversity of women's experiences and fights for equality for all, regardless of race, sexuality, or socioeconomic status.

Moreover, reflecting on this journey brings to light the critical role of men in supporting and advocating for gender equality. It challenges men to confront their own biases, to listen and learn from women's experiences, and to stand as allies in the fight against gender-based discrimination and injustice.

In essence, appreciating the essence of being a woman is a continuous journey of learning, understanding, and evolving. It's a call to action for everyone to contribute to creating a world where women are free to live authentically, pursue their dreams without fear of discrimination or violence, and celebrate their achievements without reservation. As we move forward, let us carry the lessons learned from this journey in our hearts and minds, committed to supporting and uplifting women in every aspect of life. The journey towards gender equality and the full appreciation of womanhood is one that we undertake together, shaping a future that honors and respects the invaluable contributions of women to our world.

This book serves as both a reflection on the journey of womanhood and a roadmap for the future. It's a reminder that every one of us has a role to play in supporting and celebrating women. Let's embrace this call to action with open hearts and minds, committed to making a difference in the lives of women around us and for generations to come.

Appendix

For those inspired to delve deeper into the realms of women's rights and achievements and to actively engage with organizations championing these causes, here's a curated list of resources and entities dedicated to empowering women across the globe.

Resources for Further Reading and Exploration

1. **"We Should All Be Feminists" by Chimamanda Ngozi Adichie**: A powerful essay that offers a unique definition of feminism for the 21st century, rooted in inclusion and awareness.
2. **"The Second Sex" by Simone de Beauvoir**: A foundational text in feminist literature, exploring the history of the treatment of women and laying the groundwork for future feminist thought.
3. **"A Room of One's Own" by Virginia Woolf**: An extended essay that addresses the historical lack of women's access to education and the literary world.
4. **"Half the Sky: Turning Oppression into Opportunity for Women Worldwide" by Nicholas D. Kristof and Sheryl WuDunn**: This book addresses the challenges women face worldwide and presents stories of extraordinary women overcoming adversity.
5. **"Bad Feminist" by Roxane Gay**: A collection of essays that explore the complexities of being a feminist in today's world, blending personal experience with cultural analysis.

Organizations and Initiatives Supporting Women Globally

1. **UN Women**: The United Nations entity dedicated to gender equality and the empowerment of women. UN Women supports international efforts to protect women's rights and to expand their opportunities in political, social, and economic

spheres.

2. **Global Fund for Women**: This organization is one of the world's leading foundations for gender equality, dedicated to providing funding and resources to groups led by women, girls, and trans people.

3. **HeForShe**: Initiated by UN Women, HeForShe is a solidarity campaign for the advancement of gender equality, inviting men and boys to be active agents of change alongside women and girls.

4. **Girls Not Brides**: A global partnership committed to ending child marriage and enabling girls to fulfill their potential. The organization works through advocacy, community engagement, and research.

5. **Malala Fund**: Founded by Nobel Prize laureate Malala Yousafzai, this fund is focused on advocating for the education of girls in developing countries, aiming to break the cycles of poverty and inequality.

6. **Women for Women International**: This nonprofit humanitarian organization provides practical and moral support to women survivors of war, helping them rebuild their lives through education, training, and resources.

7. **She Should Run**: Focused on inspiring women to run for office in the United States, She Should Run provides resources, mentorship, and support for women considering a future in politics.

By engaging with these resources and organizations, individuals can expand their understanding of women's rights and achievements and contribute to the global movement towards gender equality. Whether through education, advocacy, or direct action, every effort counts in the pursuit of a more equitable and just world for women and girls everywhere.

Afterword: The Author

In the role of an author deeply immersed in the study of human behavior and the diverse tapestries of women's lives across cultures, this book seeks to illuminate the rich and varied experiences of women. It's drawn from observations and interactions across different societies, recognizing that while these insights do not encapsulate every woman's experience, they offer a lens through which we can appreciate and celebrate womanhood in its many forms. This narrative invites us to be celebrants—honoring and uplifting the women around us, acknowledging their struggles, achievements, and resilience.

The book emphasizes the significant role that men can play in supporting and advocating for women. It outlines various ways men can contribute, not just as allies but as active participants in the journey toward gender equality. From engaging in open, respectful dialogues to advocating for policies that ensure women's rights and equality, men are encouraged to stand in solidarity with women.

But beyond these actions, the book calls for a collective effort, urging all readers, regardless of gender, to come together to foster a more harmonious and supportive community. It's a call to action for everyone to engage with the principles outlined in the pages—principles of empathy, respect, and mutual support. By doing so, we not only honor the women in our lives but also contribute to a world that values and celebrates the contributions of all its members.

For anyone who knows a woman who could benefit from this message—be it a friend, family member, or colleague—now is the time to act. Engage with this book and its teachings, embody the spirit of celebration and support, and become a beacon of hope. In embracing these ideas and living in the light of positivity and empowerment, we all can play a part in creating a more inclusive, equitable, and harmonious world. Let this book serve not just as a collection of observations but as a catalyst for change, inspiring each of us to contribute to a society where every woman is celebrated for her inherent worth and potential.

Don't miss out!

Visit the website below and you can sign up to receive emails whenever Jeremy B. Sims publishes a new book. There's no charge and no obligation.

https://books2read.com/r/B-A-CRSAB-EDYYC

BOOKS2READ

Connecting independent readers to independent writers.

Did you love *The Woman Celebrant*? Then you should read *What are YOU fighting for?*[1] by Jeremy B. Sims!

Discover the enemies we encounter in the modern world, from systemic injustices to misinformation, and learn how to navigate them. Explore the foundations worth fighting for—virtues like personal greatness, honesty, and truth. Acquire practical tools for positive change, resilience, and fostering goodwill.

But this book goes further; it challenges misguided battles, such as the pursuit of status and hollow opinions. It invites you to reflect on your values, beliefs, and achievable goals, empowering you to foster an inclusive environment in your life.

In the end, "What Are You Fighting For?" is a call to action, encouraging you to identify your battles and embark on a personal

---

1. https://books2read.com/u/bo9veR

2. https://books2read.com/u/bo9veR

journey of growth. It offers practical exercises, real-life stories, and resources for further exploration, enriching your path to a life filled with purpose, empathy, and understanding.

Read more at https://www.jbsims.com/.

Also by Jeremy B. Sims

Book 2
From Country Roads to City Penthouses Part 2

Standalone
Awakened Wellness: Are YOU Spiritually Motivated Yet?
Stop Blaming the Adversary: It's You!
From Milk to Meat: The Journey of Spiritual Maturity
Reviving Your Sacred Space: The Call for True Worshippers
Cultivating Self-Discipline: A Path to Personal and Universal Success
The Morning After
In the Shadow of Holiness
Un Rêve à Washington D.C.
Keep That Temple Clean: A Biblical Perspective
Harboring Harmony: Navigating Through Love, Freedom, and Moral
Rediscovery
and This is just the Beginning
50 Shades of Salvation
Embracing Authenticity
What are YOU fighting for?
Got Humility?
Check your Spam Folder!
Harmony"s Horizon

The Woman Celebrant

Watch for more at https://www.jbsims.com/.

9 798224 631162